Charles Frederick Winslow

The Cooling Globe

The Mechanics of Geology

Charles Frederick Winslow

The Cooling Globe
The Mechanics of Geology

ISBN/EAN: 9783744662352

Printed in Europe, USA, Canada, Australia, Japan

Cover: Foto ©berggeist007 / pixelio.de

More available books at **www.hansebooks.com**

THE

COOLING GLOBE;

OR,

THE MECHANICS OF GEOLOGY.

BY

C. F. WINSLOW, M.D.,

AUTHOR OF "COSMOGRAPHY," "PREPARATION OF THE EARTH FOR THE INTELLECTUAL RACES," AND OTHER DISCOURSES.

———————

" And God said, Let the waters under the heaven be *gathered together* unto one place. and let the dry land appear; and IT WAS SO." — GEN. 1. 9.

BOSTON:

WALKER, WISE, AND COMPANY,

245, WASHINGTON STREET.

1865.

BOSTON:

PRESS OF JOHN WILSON AND SON,

15. WATER STREET.

TO

HIS EXCELLENCY, JOHN A. ANDREW,

Governor of Massachusetts,

WHOSE ENLIGHTENED STATESMANSHIP — EVERYWHERE CONSPICUOUS IN THIS REMARK
ABLE ERA OF OUR NATIONAL HISTORY — HAS BEEN ESPECIALLY DISTINGUISHED
BY FORESIGHT IN FOSTERING PUBLIC EDUCATION, IN FOUNDING TEM-
PLES OF LEARNING, AND IN ENCOURAGING THE PROGRESS OF
GENERAL SCIENCE, — THE ONLY SOLID BULWARKS OF
AMERICAN CIVILIZATION AND COMMON LIBERTY,
THE VERY "ARK OF THE COVENANT"
OF OUR REPUBLIC,

The following Pages are respectfully Dedicated

BY HIS FRIEND,

THE AUTHOR.

INTRODUCTION.

THE two memoirs herewith presented to scholars and thinkers, and intended to be permanent contributions to knowledge, were read before the Boston Society of Natural History: the first, at the meeting of Oct. 5, 1859; the last, at the meeting of Jan. 4, 1865.

The geological proposition which I have attempted to demonstrate is, that *the globe, when first giving birth to life and rudimentary forms, was much larger in all its diameters than now.*

When walking in darkness, we step timidly. When the sun rises, we tread boldly. No doubts now remain in my mind upon the important subject involved in the subsequent discussions.

. When led to, and while pursuing, researches which have ended in these disquisitions, the opinions of the ancients were unknown to me. Geology, in its solid conceptions, however, is only of recent birth. It belongs to this century. It is still in infancy; and the indefatigable labor of the largest intellects will be long in bringing it to maturity. The opinions of antiquity are interesting, only as showing the sagacity of acute minds. They are valuable, because, although without scientific foundations or forethought when recorded, they were prophetic, and exhibit a clear comprehension of geographical facts, lost sight of, indeed, in this truly scientific age, in the development of geological and astronomical theories.

Plato imagined that an immense space where the Atlantic Ocean is now extended, was filled by land.

Pliny transmits the traditions of many sudden submergences of cities, lands, and mountains, and of the production of bays and gulfs, by this cause. He says, "The earth feeds upon itself,"— a crude idea, but *pointing* toward truth.

Nothing was positively known by the ancients: and nothing more was imagined, until the true scientific period, ushered in by Copernicus, began to dawn; when Leibnitz, the first to perceive that our planet originated in a molten state, inferred that, in cooling, it must necessarily shrink. He truly imagined the mechanism by which its irregularities of surface have been produced; that, in shrinking, immense caverns would be formed, into which the surface would sink, thereby forming oceans, and leaving land above water. This was in 1683. Nothing was thought of these ideas.

Deluc, in 1779, reiterated the same conjectures. But geology, like all science, needed facts for its organization into system; and, at that time, geology, as a science, did not exist. Deluc and his views were forgotten.

Within the last fifty years, scientific inquiry has been active and fruitful. The necessity felt by enlightened governments for investigating the resources of their domains has led to geological surveys, employing some able observers, and ending in solid acquisitions of knowledge. But from these acquisitions has sprung a science vitiated by the opinions of modern European geologists, that all visible lands have been *upheaved* from the bottom of the seas.

The object of the following discourses is to overthrow these opinions, to place Geology beyond theory, and to establish it, as a science, on solid foundations; in a word, to Americanize it.

GEOLOGICAL REVOLUTIONS.

.

___

READ BEFORE THE BOSTON SOCIETY OF NATURAL HISTORY,

WEDNESDAY EVENING, OCT. 5, 1859.

2

GEOLOGICAL REVOLUTIONS.

PERHAPS nothing has so much retarded the solid progress of geology as a single aphorism laid down by Sir Isaac Newton, and accepted by subsequent astronomers. He concluded, after certain mathematical computations, that the unequal diameters of the earth at its poles and equator were the result of centrifugal force exerted at the equator, in consequence of planetary rotation.

What mathematicians pronounce unalterable law, speculative philosophers and the boldest thinkers shrink from questioning.

For nearly two centuries, the present obliquity of the earth's axis to the plane of the ecliptic has been maintained by astronomers as a stable fact; and the physical results of unceasing rotation are supposed to become more pronounced and permanent from age to age. Even within three years, an eminent geologist of the United States has declared his conviction, that the North-American continent is undergoing steady geological development from the embryonic form impressed upon the earth's crust as it began to shrink by the radiation of its primal heat. More recently still, a noted Harvard professor of mathematics has announced an opinion, based on purely geographical considerations, sustaining this view ; and he expresses the conviction, on what he considers indubitable proof, that the obliquity of the earth's axis to the plane of the ecliptic has never undergone the slightest change since the dawn of creation. Indeed, the eternal fixation of the earth's axis at an inclination of 23° 28′ to the plane of its orbit seems to have become a settled astronomical idea ; and the explanation of all geological anomalies, such as the location of fossil flora and fauna of tropical species in the temperate and frigid zones, has been bent to conform to it. For instance, to account for the wide distribution of coal measures, it is believed that there was a carboniferous age of incalculable duration, during which our planet was bathed in perennial summer, from pole to pole, notwithstanding it then revolved upon its axis and around the sun in the same manner as at the present time.

Strange as it may seem, geologists have, heretofore, been satisfied with these explanations, and have bowed, without a question, to the doctrines of the astronomers and mathematicians. The views of the former have been shaped to conform to the *alleged* discoveries deduced from the demonstrations of the latter. Although naturalists have been constantly filled with wonder at the discovery of fossil organisms wholly out of place, their opinions respecting the origin of these things have been clouded, and their inquiries circumscribed, by the assertions of Newton and his followers. They have paused, as if paralyzed by reverence, at the announcements of that great philosopher; and reposed, as if the end of knowledge had been reached. They have interrogated nature for no other explanation of these climatic anomalies. Devoid, in general, of tastes for cosmical research; neglecting the study of central forces; and not comprehending the form, power, and play of agencies radiating from, or shut up, and quietly or convulsively acting within, the bowels of planets, — they have overlooked those mighty facts and possibilities which would elevate physical geography and geology to their legitimate rank among the sciences; and which, indeed, would equally contribute to advance our knowledge of astronomy.

Newton's severe mathematical turn of mind seized upon the smallest facts, and followed them to the most comprehensive conclusions. The falling of an apple led to his immortal discovery of gravitation; the twirling of a mop, to the centrifugal tendency of matter to move from the poles toward the equator of rotating planets. Thus it will be perceived, that a purely mechanical idea lies at the basis of all computations and experiments, by which it has been demonstrated that the equatorial diameter of the earth is twenty-six miles greater than its polar diameter. Experiment and observation showed a certain inequality; numbers determined its quantity. The same idea led to telescopic examination and micrometric measurements of other planets, by which it is ascertained that all are more or less spheroidal; yet Jupiter's shape, although apparently agreeing with theory, is so strangely outlined as to lead to the belief, that there is not that equable distribution of matter by centrifugal motion, which should be expected in a plastic body subject to so rapid a rotation.

Admitting the theoretical possibilities of these mechanical conditions in planetary masses (whose central forces, however, are wholly regulated by solar influences, and not subject to the accidents of centrifugal mechanics, as I have shown elsewhere, in my memoir on the " Central Relations of the Sun and the Earth "), does it follow, that the present inclination of the earth's axis to the plane of the ecliptic is permanent, and that no internal disturbances can arise to suddenly and profoundly modify this inclination ?

I have for some time ventured to entertain doubts upon these points ; and the reasons for these doubts, and for contrary opinions, are derived from the application of the simplest physical laws to this great geological problem.

An illustration will more clearly open this subject, and will more forcibly present the facts and principles upon which this inquiry is founded.

Any homogeneous and perfect sphere — a marble, for instance — will fall from a state of rest, in a vacuum, without rotating or changing its position.   But change the form of that body to a spheroid, or project mountains or sink depressions on various parts of its surface, it will change positions, partially rotating, and fall, with its heaviest hemisphere, that is, with its longest diameter, or major axis, toward the earth.   The active intellect can multiply illustrations of this sort without end ; and they involve the simplest principles of physical law, — principles which are of universal application to an atom or a world.

This planet (discarding rotation) is only a marble of enormous dimensions and heterogeneous nature, falling, in a vacuum, to the sun. Whatever may have been the cause of its primitive irregularity of form, the slightest preponderance of matter at one point of its surface or another will necessarily alter the relations of its poles to the plane of the ecliptic. The diameter of the globe at the equator is twenty-six miles greater than at the poles; and the inclination of its axis to the plane of the ecliptic is adjusted by laws of gravitation so fixed, that it can never alter, unless the relative amount of matter in the hemispheres be disturbed. Should this disturbance ever occur from any cause whatever, or in any manner, the degree of axial inclination must shift, either suddenly or slowly, according to the agency operating to produce disturbance of equilibrium. For instance, suddenly transplant Greenland from its present connection with the bottom of the Atlantic to the south of Australia, what would be the geological and the geographical changes upon the surface of the whole globe? As small as Greenland is in superficial extent, the earth would instantly feel the disturbance of its equilibrium, and the inclination of its axis to the plane of its orbit would become sensibly affected. The physical results, and changes in distribution of organic life, that would necessarily follow, would be universal in character and extent. Similar consequences would naturally follow the slow elevation or submergence of continents. This conclusion is unavoidable. But the present state of the earth's surface, the vast inequalities of its crust, embracing its varied and entire stratilogical phenomena,

clearly suggest that all its aspects are wholly independent of the mechanical results of centrifugal motion,
— an idea to which geology, with all its magnificent
contemplations, and the great truths to flow from
its boundless developments, has been enchained, since
the rocks first claimed the attention of observers and
thinkers.

Having discarded this astronomical impediment to
geognostic progress, and laid down a principle on
which paleontologists can hereafter proceed with firm-
ness and solid success, I will note a few facts, and
deduce from them the mechanism by which geological
revolutions have been produced, and through which
have resulted the present shape of our planet, and its
polar relations to the plane of the ecliptic.

The first series of facts to which I would ask attention is that where it is indisputable that the earth's
crust, in small areas, has suddenly sunk below surrounding levels. Such a spot is Kilauea, embracing a
circumference of nine miles, on the slope of Mauna
Loa, in Hawaii, where the walls of the pit are six
or eight hundred feet in perpendicular depth. This
area was ingulfed at some remote period, into the
Plutonic caverns of Hawaii; and around the ridges of
its *débris*, still projecting from the floor of the pit, the
lava rises and breaks away and falls, at long intervals,
like the tides of the ocean, which rise and freeze around
the coast of other half-sunken islands.

A similar spot is Thingvalla, the seat of the ancient
Althing, in Iceland, well described of late by young
Lord Dufferin. This, however, is greater in extent

than that of Kilauea; and, happening to be nearer the coast, it became partly overwhelmed by the sea.

A portion of the site of old Callao is said to have sunk in an instant, on the 28th of October, 1746. Ships now anchor above its dwellings, and mephitic gas is frequently ejected from the bottom of the bay.

The quay of Lisbon was ingulfed, with equal suddenness, on the 1st of November, 1755 ; and a large area near New Madrid, in the Valley of the Mississippi, vanished, in the midst of a terrible earthquake, on the 16th of December, 1811, and was overwhelmed by the floods of the river.

These submergences, preceded by terrific subterranean thunders and convulsions, disclose the remarkable geological fact of cavities beneath, of great depth, whose dimensions must be greater than the superincumbent masses received into them. These statements conclusively settle the question of subterranean cavities on a small scale ; that is, from two to twenty and thirty miles in extent, into which the superincumbent crust may plunge, and be overwhelmed with water, or with the plastic matter of the nucleus. Records of phenomena of this kind are very numerous.

From these minor facts, it is easy to ascend to greater, and show the same relation of superficies to internal cavities and canals, by the evidence of strata of identical character, exposed on opposite coasts, headlands, and many river channels, in consequence of the submergence of intervening areas of the crust. The great lakes of North America, the Mediterranean Sea, the German Ocean, and the English Channel, are but larger expressions of the same geological events. The limestones of Cuba and Yucatan are but continuations of the former sea-bottoms of North America; the sudden collapsing of which upon the condensing nucleus of the planet gave origin to the Gulf of Mexico, and Caribbean Sea, with its linear groups of isolated fragments, and to the southerly dip of this northern continent, — in the midst of which catastrophe, as the globe swayed from its former equipoise, the oceans swept to the new equator, and perhaps effected that remarkable drift action so conspicuous in our hemisphere.

From these illustrations, we may ascend to still more grand displays of similar submergences, where the evidence is clear and positive, although different from that heretofore presented. I will first allude to the ingulfment of a continent that once occupied the present area of the Indian Ocean, the scanty vestiges of which manifestly disclose its former existence. Among these vestiges is the remarkable island of St. Paul's, which I have personally explored. It is the top of some vast volcanic mountain, like Cotopaxi, or Popocatapetl, which lingers above the waves as did the chimneys of some of the houses of old Callao, for a while after its submergence. One side of the crater was split off in the catastrophe, and sank many hundred feet deeper than the existing island, admitting the sea more than a hundred feet deep within its basin; while a needle of immense dimensions stands out from the main land, an unmistakable sign of the nature of the accident, and contradicting all possibility of upheaving action.

So, in the broad Pacific, the long chains of atolls, as ingeniously divined by Darwin, plainly betoken the existence of submerged mountain-ridges, and of craters, which, in former ages, played an important part in throwing off the heat of the shrinking nucleus, and preparing the globe for the destructive cataclysms which have divided the geological ages. The evidence of sunken craters, derived from lagoon formations, is strengthened by the half-sunken crater of St. Paul's, in the Indian Ocean, just described; while, the coral chains of the Laccadives and Maldives, further north, are only a repetition of the same phenomena, much more largely distributed, in the tropical latitudes of the Pacific.

Partial subsidences of great areas are as palpable as total submergences. The general dip southward of North America is a remarkable example of this; and minor ones are as numerous as the water-sheds and synclinal valleys throughout the planet.

But the question will naturally be asked in this stage of our inquiry, What substantial evidence now exists of empty spaces beneath the crust? — what more than the mere presumption of their former existence derived from appearances of Kilauea, the great lakes, British Channel, and other illustrations?

The reply to this question immediately brings into systematic requisition an important series of facts heretofore neglected by physicists; those remarkable subterranean thunders long observed, many of which are noted in Humboldt's writings, and have been heard by my informants in the mountains of California, and by myself in the western part of Mexico, unaccompanied with the slightest perceptible motion of the ground.* I need hardly refer to the records of science to recall the terrible thunderings, which, on the 12th of April, 1812, suddenly startled the inhabitants of Venezuela, and were heard over an area of ninety-two hundred square miles; while over six hundred miles distant, in the northeast, the volcano of St. Vincent poured forth a copious stream of lava. By this phenomenon, we may safely declare, in view of the present state of our knowledge, that the eruption of molten rock into a vast cavern, or chain of caverns, deep down below the bed of the Caribbean Sea, found its way to the surface, among the West-India Islands, decomposing water or atmospheric air, and igniting explosive gases in its passage; which reverberated through, and plainly disclosed, the existence of enormous voids underlying the northern and eastern portions of South America.

A similar phenomenon occurred, in 1744, in New Grenada, where subterranean sounds resembling the discharge of cannon were heard over a large area;

---

* Since the composition of this discourse, the author has extensively travelled in the Andes and the coast-regions of South America, where he has observed these phenomena repeatedly. In the Andes, at Riobamba, the terrific subterranean thunderings were very frequent, and seemed to proceed from the direction of the violently active volcano of Sangai, over thirty miles distant. These profound bellowings, rolling along for indefinable distances, and reverberating with continuous rumblings, were generally unaccompanied with the slightest motion. Sometimes, however, there was sensible quivering of the ground, which I have felt in the stillness of night, while lying in my bed in Riobamba, and when I have been travelling during the day. I have observed the same phenomena, also, both west and north of Cotapaxi, when I supposed they proceeded from the subterranean caverns extending beyond the flanks of this active volcano.

Similar phenomena I have observed often at Guayaquil, where the sounds and motions appeared to come from the direction of the Pacific, following up the Gulf, and proceeding eastward towards the Andes. I have felt the jars of these subterranean detonations, while on board a steamer lying in the river.

while Cotopaxi, eighteen thousand feet higher, and four hundred and thirty-six miles distant, sent forth a violent eruption. Here is another suggestive fact, like the preceding one, where molten floods, moved by repulsive forces, were elevated nineteen thousand feet above the level of the sea, and perhaps three hundred thousand feet from their source; and, in their passage from the central fires, opened communication with immense subterranean cavities, exploding gases which resounded several hundred miles, beneath the northern and western part of South America.

During the violent earthquake of New Grenada, in February, 1835, subterranean thunders were heard simultaneously at Popayan, Bogota, Santa Martha, Hayti, Jamaica, and on Lake Nicaragua. At Caraccas, they continued seven hours, without any movement of the ground. These facts, with various others, have been collated by Humboldt; and in the last, as in the other two, the same phenomenon is exhibited, indicating an area of 14° of latitude by 25° of longitude, embracing a large portion of Central America, the West-India Islands, and the whole northern part of South America, — an area of one million and fifty thousand square miles, to be more or less undermined, and *liable to sudden submergence.*

All these facts — to say nothing of others similar or varied in their expression, and indicating even broader continental ranges exposed to future subsidence — are solid evidences, so far as they can be traced, and applied in all their bearings, to substantiate the geological position laid down, and the conclusions maintained in this discourse. While they lay foundations for the broadest discussions, they elevate geology into its true position among the sciences, by opening a new field to the philosopher, stored with the boundless wealth of past time. The *modus agendi* of subsidences has been heretofore only matter of conjecture, and an unsolved problem in the physics of the globe. Leibnitz — that greater than Newton, in the general amplitude of his understanding and acquirements — was the first to suggest the original igneous condition of the planet, and its gradual shrinkage from the persistent radiation of its heat. Cordier, more recently, has reiterated this opinion, and fancied that it must, as a matter of course, become wrinkled, like a drying apple, in undergoing this process; and that thus the strata have been slowly folded, the continents and mountains uplifted, and oceans formed. Geologists generally have settled down upon this theory of upheavals; and James D. Dana, the most accomplished scientist in this country, has endeavored to systematize the idea, by an ingenious disquisition on " Geological Development," which is radically defective by reason of its inapplicability to universal conditions. But the illustrations which I have presented, and the facts upon which I demonstrate the *modus agendi* of sudden ingulfments of vast areas, while they 'exhibit the penetrating sagacity of Leibnitz respecting the effects of radiation of heat, disprove the theory of Cordier, and the speculative developments of Dana; and present geology to scholars in such aspects, and based on such principles, that there is no problem, however obscure heretofore, that does not receive a ready solution among its deductions.

By these sudden ingulfments of great areas, the equipoise of the world would be disturbed, the inclination of its axis to the plane of its orbit would be modified; the oceans would instantly change their beds, leaving myriads of aquatic animals on dry land, or embedded in drifting sediments, and exposing water-lines on the margins of rivers, lakes, hills, and islands. Terrestrial animals would be suddenly transferred, perhaps swept by floods from temperate to frigid regions. New exposures to the solar rays would supervene, profoundly affecting the isothermals, so that the glaciers of the poles would be dislodged, and set in motion over broad regions; and the theory of moving ice, suggested by a gifted naturalist, which heretofore has been so embarrassing to geologists, would be transformed into a living and immortal truth.

From this point of view, also, the more profound secrets of the globe will become better understood; such as the penetration of the ocean into seething caverns, which, becoming vast reservoirs, and retorts for the solution of silex and metals, collapse again in the process of time, and press up their contents through fissures, in the shape of metalliferous-quartz veins; such also as the origin and phenomena of certain earthquakes, and of subterranean thunders, of salt mines, of immense petroleum deposits, and, among other things, the unnatural warmth of the gulf-stream, by the return of heated waters from sub-oceanic fissures in the Bay of Mexico.

It does not absolutely follow, because planets are spheroidal, that their oblateness springs from the cen-

trifugal effects of rotation. It is an established law,
that all bodies fall with their major axes, or heaviest
diameters, toward their centre of gravity; and since it
is now unquestionably ascertained that causes lying
within the bowels of the earth, and wholly independent of the action of centrifugal force, are periodically
operative to change its shape, and modify the inclination
of its axis to the plane of its orbit, it becomes an object
of interest to geologists to inquire if the earth's diameter at the poles was not even greater when organic life
was first created than its present equatorial measure-
ment. It is reasonable to believe that it must have
been so; for, when all the greater and minor changes
which can be readily traced are contemplated in the aggregate, I must confess my conviction, *that the planet has
diminished many miles*, IN ALL ITS DIAMETERS, *since organic forms were first introduced into its seas;* and, if
so, it becomes self-evident, that those seas, from being
very shallow, and of universal distribution in the primal
ages, have become deeper, and divided by ever-changing
coasts, sweeping, with every modification of the earth's
equipoise, over old lands, and collecting into new basins,
until their present geographical limits were defined by
the last great catastrophe, which, judging from recent
discoveries, must have occurred since the creation of the
human race. These events must be necessary consequences of the physical constitution of our planet, and
of the action of its central forces; and I have little
doubt that the further pursuit of this inquiry will lead
to a certain sort of chronological accuracy respecting
those grand dynastic revolutions, which, in the his-

tory of the globe, have divided its geological epochs, when new conditions of climate, surface, of magnetic and organic forces, springing from its successively new and nearer relations to the sun, combined for the production of new and more elaborate forms of vegetable and animal life. So numerous are the evidences of vast continental revolutions that it is physically impossible that the equipoise of the planet should not have been many times profoundly disturbed ; and, if so, the *present* inclination of its axis to the plane of the ecliptic must be an event of comparatively recent occurrence. The evidences derived from earlier geological conditions leave no doubt of the earth's former polar, or almost polar, relations to the sun's vertical beams.

The various inclinations of the axes of the other planets to the plane of the ecliptic strongly substantiate these deductions. The extreme difference of inclination in that of Jupiter, and in that of Uranus, — the first, perpendicular ; the last, almost transverse, — demonstrates causes at work, which, inferring from what we know of the constitution, forces, and revolutions of our own globe, must proceed from the centre of their respective spheres, We find, indeed, the moons of Uranus pursuing retrograde paths ; thus pointing us to the wondrous fact, that his revolutions of surface have been so vast or numerous, as to completely reverse the ordinary and original direction of the poles ; so much so, indeed, as to produce a retrograde rotation, which his satellites, by well-known cosmical laws, must necessarily follow. The retrograde motions of the satellites show a remarkable local peculiarity : but, by geological

4

deductions, this anomaly is satisfactorily explained, and a uniformity of design, origin, and construction through-out the system, is established.

The pursuit of this wide geological inquiry opens into the realms of astronomy; and we behold the sub-limest of sciences becoming tributary to the rocks for the explanation of its most difficult problems.

The inequalities of surface in the heavenly bodies nearest the earth show the same causes at work within their bowels, which, acting within this globe, have pro-duced the countless disturbances of its rocks and oceans. Change the relative bulk of matter in either hemisphere, and the planet would sway in space, and turn its longest diameter to the sun. New poles and a new equator would be established. The degree of axial inclination to the plane of the ecliptic would become greater or less; and according to the extent and suddenness of these changes of equipoise would be the movements of the seas, the exposure of reefs and shallows, and the over-whelming of old coasts. The climate everywhere would be changed, and the destruction of entire genera of plants and animals would ensue; or, such profound alteration of terrestrial forces be effected, that former ranges of distribution would be annihilated, and old forms would take on new expressions of development, until, at last, as in the present epoch, the subtlest principles of na-ture would become elaborated into a frame-work of atoms, — a compound of all the past in plan, substance, form, force, and instinct, — with an internal fitness added for companionship with the great creative Thought.

RADIATION AND GRAVITATION

APPLIED TO

GEOLOGY AND PHYSICAL GEOGRAPHY.

READ BEFORE THE BOSTON SOCIETY OF NATURAL HISTORY,

WEDNESDAY EVENING, JAN. 4, 1865.

PREFACE.

RADIATION, as a *cosmical* fact of immense and universal significancy, was justly the discovery of Leibnitz. GRAVITATION was equally the discovery of both Newton and Leibnitz. While the latter agent has been allowed its real value, as a force of universal and creative activity, and has elevated astronomy to the loftiest ranges of exact study; the former has been employed only in a speculative sense; and, so far, has led theorists to no knowledge of its truly practical and effective agency, as a co-worker with gravitation, in the present economy of the physical universe. The nebular theory in astronomy, and the upheaval theories in geology, have exhausted the wits of their advocates; although physical facts are known, which, if candidly weighed, would reduce all these speculations to absurdity.

The time has arrived for ingenious theories, however renowned their authorship, to yield to the commanding voice of inexorable nature.

The effects produced upon our planet by the radiation of its heat are only counterparts of circumstances transpiring in all planets, and in all cosmical masses,

however vast their dimensions, or however remotely separated from each other. While gravitation pervades all nature, and, however atomic in its accommodations, is, nevertheless, a strictly astronomical and world-sustaining agent; radiation, more local, assumes cosmological functions, and becomes the world-maker and world-perfecter. As will be seen in the following discourse, it prepares the globe for those sudden and terrific revolutions which end in preparing its surface for new forms and races of plants and animals. Studied from this point of view, it becomes magnified into a great visible and practical fact, no longer to be overlooked by geologists. It is the veritable fulcrum of their science; and is as important to the progress of their department of knowledge as gravitation has been to the developments of astronomy. Both are equally effective forces in the hand of God. Both were the handmaidens of his power, when Chaos sprang to light and life. Both are the angels of his love and wrath. Both silently obey his mandates in the evolution of progressive good.

c

RADIATION AND GRAVITATION

APPLIED TO

GEOLOGY AND PHYSICAL GEOGRAPHY.

I HAD the honor, on the 5th October, 1859, to read a
memoir before this Society, advocating a system of terrestrial changes different from the current theories of
geology. It attracted no attention, received no notice,
and excited no discussion. The views then advanced
were the result of much reflection upon the structure of the earth, after extensive travel, and vigilant
observation of its surface. Larger opportunities for
investigation have only confirmed the general views
at that time initiated ; and I now, respectfully and
earnestly, invite the attention of geologists and physicists to that subject, as I shall herein present it, and
solicit them to examine it in all its aspects and developments.

Paul was brought up at the feet of Gamaliel, and
believed in the straitest doctrines of the Jews. But
questionings at last fell upon him as he journeyed ; and,
after groping awhile in darkness, his eyes suddenly
opened to marvellous light and truth. Geology has
its Gamaliels ; and we have all been brought up at

their feet. The doctrine of upheavals by slow or active igneous processes is an established idea among all scholars and theorists ; and it is as firmly fixed in the minds of geologists as were the law and prophets in the minds of the Hebrews.

The general irregularities of the earth's surface ; the broken and dislocated conditions of rocks everywhere, from the highest mountain summit to the lowest fossiliferous strata and the bottom of the deepest mines ; the presence of marine forms, high above, and far removed from, their ocean homes, and of coal - fields full of tropical plants, under the earth and sea, and in arctic regions, — have, from the days of earliest knowledge, excited speculative curiosity, and, beginning with Leibnitz, have stimulated thinkers to attempt solutions of these wonderful problems.

The declarations and authority of two notable men, more than the statements of all others, have retarded the advancement of general knowledge. The personages to whom I allude are Moses and Newton. The authority of the former weighs, even now, too much with observers and natural philosophers ; and I am free to say, that the severest truth, as it lies in nature, cannot be faithfully sought nor found, until we rise above all the sacred traditions of the ancients. Indeed, it is only after throwing off all shackles of church and school, and striking the rock with our own inspired rod, that we can obtain that living truth which shall water the earth with its unfailing fountains.

Whatever may have been the retarding influence of the Jewish legends upon the free expansion of the human intellect, nothing has more hindered the development of cosmographical truth than the mathematical authority of Isaac Newton. The discoveries of Copernicus (born 1473), Tycho Brahe (1546), Galileo (1564), and Kepler (1571), opened modern science for the application of the mathematics. A century later arose Newton and Leibnitz, born about the same time, the first in 1642, the last in 1646, whose studies embraced all the departments of physical nature. Peers in all the elements of intellectual greatness, endowed with similar gifts, equally discoverers of gravitation and of fluxions independently of each other, they appear to have been sent by Providence for the highest special purposes; and their names must forever remain the proudest in the history of the exact sciences. The researches of Leibnitz convinced him that the earth, at some former period, had been in a state of igneous fusion. Its oblateness having been ascertained, Newton made computations by his new methods of notation, and announced that the centrifugal force incident to the rotation of the planet was just sufficient to produce the difference between its diameters at the poles and equator. This difference is now known to be not far from twenty-six miles. All experiments and observations demonstrate a certain inequality; but it is now known, that the oblateness of the opposite hemispheres is dissimilar.

The mathematical deduction of Newton has so influenced the opinion of physicists, and the developments of geology, that while all appearances of the earth's crust indicate numerous, successive, universal, and sud-

den changes of climate, and of the aqueous and solid
materials which compose and envelop it, no man has
ventured to question, or attempted to disprove, the accuracy of his demonstrations and conclusions.   They
remain settled doctrines in the physics of the globe to
this hour.   Geological theories have been narrowed and
moulded by them ; and the alleged *experimentum crucis*
announced by the illustrious author of the "Principia"
has given them the appearance of infallibility.

All irregularities of the earth's surface are studied
from this point of view.   It is declared by astronomers,
that the relation of the axis of the globe to the plane of
its orbit has never changed since its creation, except,
possibly, by slow and returning librations from tropic to
tropic.   Astronomy and the mathematics are the oldest
sciences ; and have been respected as the most sublime
and exact of all the departments of physical learning.
Geology, of later birth, has held itself humbly sub-
ordinate to them in all its aspects and speculations.
Astronomers have promulgated planetary laws, declar-
ing them invariable ; while geologists have contented
themselves with simple observations of rocks and fossils,
and striven to harmonize their doubts with the possible
errors of their lawgivers.   Astronomers and mathematicians having declared the movements and shape of
planets unchangeable, geologists have quietly accepted
the belief, that the present size of our globe has been
always the same.   While conscious of many violent
changes upon its surface, they have endeavored to
reconcile their observations with astronomical theories,
and have decided that all geographical and geological

phenomena are results of slow and insensible upheavals and depressions occupying indefinite periods of time; that an upheaving mechanism within the globe is the active agent in producing all visible distributions of land and water and dislocated strata; that, while one continent comes up, another goes down; that these changes are alternate; and that all move in harmony with some general mechanical law which controls the rotation of the earth, swells the equator, flattens the poles, and maintains its axis for ever with the same inclination to the plane of the ecliptic.

Such is the doctrine, thus expressed in general terms, in which we are educated, and wherein I was myself orthodox, until my eyes were opened as I journeyed from country to country by sea and land. Thus formerly believing, and desirous to account for these phenomena of upheavals; knowing other planets to be irregular in form also, and supposing all cosmographical changes and dynamical movements might depend on some unit of force, I, for many years, labored to reduce phenomena observable in comets, in the movements of the heavenly bodies, and in the movements of the surfaces of planets, to philosophical system. In the course of those investigations, I discovered and demonstrated the existence of a *repulsive* force, co-extensive, co-eternal, co-existent, and co-equal with the force of gravitation; and published the results of my inquiries in March, 1853. The suggestion of such a force was an innovation upon the accepted doctrines of astronomy. As some know, my humble treatise remained unnoticed for several years; and, at last, was reviled, in 1857, by Benjamin Pierce, a famous professor of mathematics, and by Thomas Hill, now President of Harvard University, whom I am proud to honor as a profound and modest geometer, without a living peer.

It was declared, that the existence of such a cosmical force was not necessary to assist mathematicians to a knowledge of celestial mechanics, and that THEREFORE *it did not exist;* that the assertion of such a force vitiated literature, was superfluous in science, and embarrassing to the simplicity so apparent in the nebular hypothesis, and in the general system of nature, as planned, created, and sustained by God. These, perhaps were powerful arguments; but, fortunately for science and for truth, "The *Heavens* declare the glory of God, and the firmament showeth his handiwork; day unto day uttereth speech, and night unto night showeth knowledge:" and the most dogmatic schoolmen, when convinced, bow to this evidence with alacrity; for on the following year, 1858, ever memorable in the annals of astronomy, appeared that wonderful comet discovered by Donatti, and which is distinguished by his name. The great equatorial telescope of Cambridge revealed to the elder Bond the *positive existence,* in that body, of the cosmical force which I had already demonstrated. But, to make its existence the more certain, Professor Pierce applied the *experimentum crucis* of his mathematics to Mr. Bond's observations, and, as the result of his analysis, declared repulsion to be, not only a universal cosmical force acting in harmony with gravitation, but that he was its *original* discoverer, in like manner as Newton was the discoverer of gravitation. Lest any may suppose this curious statement an exaggeration, the fact, over his own signature, may be found printed in the "Boston Courier," of Thursday, Nov. 11, 1858.

I refer to this subject in this connection, only because it is involved in the history of my own special pursuits, and in the developments of this discourse, leaving facts and dates to the annals of American science, which will finally be as severe as fate in settling the merits of all men, when their standing, ambition, and controversies become nothing and avail nothing.

Being satisfied with the application of my hypothesis to the movements and phenomena of cosmical bodies, and having proven that earthquakes and volcanic eruptions hold inverse numerical relations to the length and sweep of the radius vector, depending therefore upon solar causation, I, nevertheless, have not been able to reconcile many geographical and geological anomalies observed by myself, with the theory of *upheavals as a result* of cosmical repulsion. I therefore resolved to study that subject more carefully, and to travel over new fields to confirm or disprove my former views. Those views, borrowed entirely from the geologists, have undergone a radical change. The doctrine which I initiated in my memoir on "Geological Revolutions," read before this society on the 5th of October, 1859, I shall now attempt to establish by a more thorough treatment.

Plainly stated, the propositions are these: That the irregularities of the earth's surface (and of all planets), i.e., our physical geography, are the results of sudden depressions and ingulfments by which the globe has, from time to time, been absolutely reduced in size, and violently careened, in different directions, toward the sun; that these events and consequent cataclysms have occurred many times since vegetable and animal life appeared in its primeval seas, rendering it probable, that the earth, when first giving birth to life, was from two hundred to three hundred miles larger than now in all its diameters; that upheavals, while they really do exist, are exceptional and limited, as in the case of volcanoes, dikes of igneous material, earthquakes, and slight oscillations of coast, — all which are attempts of the repulsive force to upheave, and resolve matter back to its elementary diffusion, counteracted and rendered abortive by gravitation.

Such are the principal points of this thesis; proofs to sustain which, will be succinctly presented in the subsequent disquisition.

Gravitation is admitted by all astronomers and physicists as an established fact and universal law; and is beyond discussion. It controls the position of all bodies upon, and suspended around, this planet, and establishes their equilibrium according to the major amounts of matter in their different diameters; the major diameters always tending toward its centre. It may be illustrated in countless ways by the simple process of weighing bodies of angular and spheroidal forms. Bodies, whatever their form, size, or weight, when suspended, and free to move in all directions, and settle into equilibrium, will finally assume that position toward the centre of the earth, which coincides with the greatest number of particles; in other words, with the greatest density or weight of that diameter which is immediately perpendicular to the surface, and in a line with the radius between it and the earth's centre. Let any given body of any size or form or weight be thus suspended, adjusted, and tranquil in equilibrium; if undisturbed by *internal* or *external* action, it will remain for ever at rest. Let any force *within* its surface disturb the relations of its various parts, its original equilibrium will be destroyed; and it will immediately change its adjustments to the centre of the globe. Let its form be changed by any *external* action, either by cutting off a portion; or, if it be a malleable body, by hammering; if an iceberg, by melting: the effect will be the same, and instantaneous or slow, according to the application of external force. A thousand acts and experiments of this sort, observable in every-day life, will serve to illustrate the operation of this common law. All matter and masses within the range of the earth's attraction are bound to its centre by this method of action of this law. It is the *operation of gravitation* in producing geological changes, which I propose to unfold in proceeding with the sequel.

That we may study its action in a wider sphere, where it begins to assume cosmographical developments, let us, for a moment, discuss the moon's relation to our planet. We will discard, in this connection, all discussion and speculation respecting the causes of rotation of the heavenly bodies, whether they arise from internal forces, or external impulses communicated by nebular condensations. The moon, as all know, presents one of its hemispheres only, to the inhabitants of the earth, and revolves on its axis in the same time that it performs its synodic revolutions. Astronomers consider this " a very singular feature." They declare that it belongs to all satellites of all planets, that its cause is unknown, and that "It is one of those dark points which stir us to profounder inquiry concerning the scheme in which we are." I quote the language of a distinguished European professor of astronomy. Now, if we apply the common law of gravitation to this singular feature in our satellite, untrammelled by theory of any sort, the problem will solve itself with extreme simplicity, and, at the same time, illustrate the application of the same law to still wider fields of research. The moon, suspended in space, and in a comparative vacuum above the earth, yields in all its parts to gravitation, and settles in equilibrium toward the earth's centre; just as an orange will gravitate, or a pear, an anvil, a huge bowlder, or a loaded ship, or an iceberg in the ocean. Orange, pear, anvil, bowlder, loaded ship, iceberg, and moon all follow the same law; and those of their diameters which contain the most matter, in other words, the greatest weight, assume that position towards the earth's centre, which gravitation demands by its inexorable force. Unlike the other objects, however, the moon moves around the earth, free, in every point of its orbit, to sway toward the earth's centre in such a way, that its heaviest or major diameter shall always remain perpendicular to the earth's surface; and the axis of that diameter will always exist as a projection of the radii of the earth over which it moves in its synodic revolutions. This is self-evident, and cannot be otherwise, if gravitation be a universal truth.

Of course, this demonstration forcibly illustrates the existence of a great cosmical law of repulsion, now admitted by astronomers, which holds the satellite at definite distances from, while gravitation fixes its relative position to, the earth. Both acting together define its orbit; while the moon itself, as a simple object suspended in space and weighed by the earth, turns its heaviest axis toward it, like any other mass of matter of different form, size, or composition.

Thus is explained that anomalous feature in the moon's motion, or rather in its state of rest, which has heretofore been an inexplicable problem, in consequence of its being entangled with projectile and nebular hypotheses, and not being considered a simple body, subject throughout its structure to the law of gravitation.

Now, as gravitation is a central force, and is exerted by the sun upon the earth, in like manner as the earth acts upon the orange, pear, anvil, ship, iceberg, and the moon, let us permit ourselves as geologists to boldly study the earth from material, rather than from mathematical and astronomical points of view.

Discarding, as in the case of the moon, all discussions relative to the *causes* of rotation in the heavenly bodies, we will consider them as simple masses of matter, with gravitating relations to each other. The larger masses always control the movements of the smaller; and all act upon each other in the directions of their largest or heaviest diameters. Thus you see nearly all the planets, whatever their distances in the remoteness of space, confined in the narrow compass of the zodiac; and every one of them moving toward and around the most ponderable diameter of the sun. Eclipses, transits, oppositions, and conjunctions constantly occurring in our solar

system, are sufficient proofs of this. And, if this point
of philosophical deduction be extended into the broadest
cosmical expansions, it will appear plain to the profound
thinker, that, while the sun causes the earth's heaviest
diameter to preponderate towards its own heaviest diameter, the same circumstance must also indicate the region
of space where those masses exist, which impress upon
the sun *its* spheroidal shape, and control *its* equatorial
motions. As the object of this paper, however, is not to
dilate upon cosmical physics, nor indulge in speculations
of any sort, but only to show the relations of celestial
dynamics to geography and geology in their physical
aspect, I will draw and confine attention to our own
planet, and to the changes that transpire upon its
surface.

The equinoctial plane is inclined to the plane of the
ecliptic, to the extent of 23° 28′; but the general fulness of the equatorial regions of the earth is such, that
astronomers declare it subject, through foreign influences,
to periodical librations from tropic to tropic, embracing
cycles of many thousand years. Some have supposed
that these alternate and insensible oscillations would account for the many changes which have evidently taken
place in the sea's level at different epochs. Geologists
of practical acquirements cannot perceive, in this assumption, sufficient cause to account for many extraordi-
nary facts within the range of their observation. But
this statement of mathematical astronomers demonstrates the general truth, that the equatorial regions of the
planet are the most ponderable; and that its equilibrium
is sensitively affected by perturbing causes. Now, it

matters not, as a necessary consequence, how causes
operate upon this globe to produce disturbances in its
equipoise.   They may act from without.   They may act
from within.   Causes acting externally lie in the do-
main of astronomy, and require no attention, especially
since such causes have been shown, in my published
discussions of earthquake and volcanic phenomena, to
end in displays of force, which appear to be re-actionary
from the centre to the periphery of cosmical bodies.   If
discussed at all, they will only add weight to the sequel
and burden of this discourse.   Causes acting within
the crusts of planets, and their results, become, then, the
legitimate and exclusive points for consideration.   Astronomy and mathematics are left behind.   Our study
becomes purely cosmographical; and, confined to this
globe, brings us face to face with the puzzling problems
of physical geography and geology.

How were the irregularities of the surface of the
earth produced?   How came undecomposed and entire
mammoths enveloped in ice near the north pole?   How
came fossil tropical vegetation of more remote ages
buried and cropping out in arctic regions?   How came
conglomerates, hundreds of feet in thickness, and scores
of miles in area, massed together, without regard
to size or arrangement of material?   How came the
rocky structure of the globe, and geological deposits
of all ages, broken up and dislocated, inverted and
slanted in all directions?   Every geological age has had
its drifts and its overturns.   What agency or mechanism
produced them?   How came the remains of extinct ele-
phants ten thousand feet above the ocean, and five hun-

dred feet deep in silt, in one hemisphere; while, at the antipodes, they are below the sea? How came mountain-chains and old volcanoes submerged and quenched in the fathomless depths of the Pacific and Indian Oceans? How came old corals, and marine forms of endless sorts, in the heart of every continent?

Let us interrogate,—not Moses nor the prophets, not Newton nor the geometers,—but the earth, and let it answer.

The ancients supposed the earth unchangeable, and that it had eternally existed as it appears to us at the present time. Leibnitz *first* comprehended the evidence that the earth began to exist in a state of igneous fusion, and that life must have had a comparatively recent beginning upon its surface. He further conjectured, that its radiation of heat had produced contraction of its mass. From his time to Cuvier, and even to our own time, a succession of physicists and naturalists have stumbled on through the mazes of discovery, puzzled by their own observations, doubting their own senses, and, naturally enough, theorizing upon the problems encountered at every step. Every great philosopher has astonished his age by his new announcements, and still truth has been only progressive. In the activity and confusion of inquiry and speculation, important landmarks in science have been sometimes lost and forgotten. Indeed, in the course of researches, since I began to write this paper, I find that Leibnitz, in his studio, imagined and divined, in 1683, what my own observations by travel have compelled me, for several years, to suspect and search after, and at last conclusively believe,—that

vast caverns exist beneath the earth's crust, the roofs of
which falling in make large basins which receive the
seas, and that what remains above the water becomes
dry land and mountains. I find, also, that Deluc, a sagacious observer and profound thinker, who studied the
Alps and Jura, published similar views in 1779, a century later. Deluc said, " Ancient continents have sub-
sided, the sea, overflowing this depressed space, has left its
ancient bed dried up, which forms our continents." But
subsequent theorists, adopting the mathematical conclusion of Newton respecting the earth as a spheroid, and
desirous to harmonize all discoveries in physics and nature
with the sacred cosmogony of Moses, have conceived that
the internal fires are a mechanism perpetually at work
to upheave the crust, while depressions are only consequences ; or, that one follows the other so as to allow the
earth to maintain the same oblate figure and general size.
Notwithstanding the accumulations of knowledge, a century after Leibnitz, it is said, by an authority no less
eminent than Flourens, that Deluc did " not perceive the
true *mechanism* which is the *upheaving* of mountains ;"
that " he stops at the *apparent mechanism*, which is the
subsidence of plains." It is possible that Deluc did not
understand the re-action of the central fluid mass upon
the crust of the planet; and I think it questionable
whether it is yet perfectly understood, although I was
once orthodox in the present ruling faith. In the days
of Leibnitz, nothing was known of geology, or of the
data which compose existing science : even in the
time of Deluc, but little progress had been made in this
department of knowledge. It is only of late that geolo-

gy has crystallized into solid system. It now invokes truth, wherever it will explain her problems, and rejects error, wherever it palpably appears. Still, the opinions of Leibnitz, respecting the radiation of heat from this cooling globe, have been applied by Cordier to a theory of condensation, which, more lately, has been explained by Dana, as resulting in corrugations, or puckerings of the crust, as the shrinking of an apple puckers its skin; by which process, continents and mountain-systems are gradually developed, and elevated from the bosom of the waters. In all the theories of upheaval, however, geologists admit (because the fact cannot be rejected) that violent movements of the seas, and violent convulsions of the whole world, have, at some time or other, supervened.

Thus the discussion culminates upon the great admitted facts, that the oceans have, from time to time, violently and suddenly moved, and that the planet has been also profoundly convulsed, and its surface overturned. How can the surface of the globe be violently overturned? How can the oceans be suddenly and violently moved? Isaac Newton and his followers cannot, necessarily, believe in any violent overturns, nor sudden movements of the waters. Geologists, however, have unequivocal evidence that both these phenomena have occurred at various times. Isaac Newton and the astronomers say, the earth has always maintained the same obliquity to the plane of the ecliptic, and that alternate summer and winter have prevailed in the northern and southern hemispheres from the beginning of time. Geologists, supposing these opinions of astronomers to be infallible, and knowing that tropical climates have

existed at the north pole, and that glacial cold has prevailed in what are now temperate zones, have indulged
the unnatural error, that the earth, enjoying universal
warmth, once bloomed in universal verdure ; and that,
subsequently and quite lately, a climate so intensely cold
has prevailed, that our temperate latitudes have been
clothed with ice full fifteen thousand feet in thickness.
Astronomers and physicists must laugh at such fancies of
the geologists ; and geologists must doubt their own reason
also, when all know, that if the earth has maintained its
present axial relations to the sun, even admitting its
librations, its summers and winters, its heats and its cold,
would always be nearly the same. The assumption of the
astronomers sets at nought the presumption of the geologists. We must, therefore, seek other causes; and if,
in the severity of our inquiry, we find reasons to doubt
the positions of both, we must stand upon the facts, and
let the mathematicians change their bases of calculation.
Geologists will welcome any theory which can solve their
problems, reduce their observations into system, and
bring their science into harmonious relations with
natural laws.

When geologists abandon all hypotheses, and rely
upon the earth alone for evidence, the fact becomes apparent, that the globe has never been stable, long at a
time, in its equatorial aspects toward the sun. It has,
beyond all question, as my observations and reasoning
teach me, been subject to sudden and frequent changes
of equipoise. It hangs in a vacuum, and is more sensitive to disturbances of equilibrium than any thing upon
its surface by which we may illustrate the action of

gravitation. For instance, a ship is loaded for sea, and her drafts are measured. If you shift a ton of her cargo, or her anchor, from bow to stern, or from side to side, she inclines according to the change of weight. But the facility of the ship's motion is impeded by the friction of the medium in which she rests. It is unnecessary to multiply illustrations upon this point. The earth sails in a vacuum, and is not hindered, in any of its movements, by any resistance ; and its *surface* holds the same relations of gravity towards the sun that a ship holds to the centre of the globe while floating upon a calm ocean. Even admitting the doctrine of *upheavals of continents* by indirect action of the sun or otherwise, the effect would be the same upon its equilibrium, although not so great as to admit that the globe, when it first brought forth life, was much larger, and that immense and sudden depressions had taken place, into which the oceans had rushed, leaving new land and mountains on a new earth.

Now, since the earth is a sphere of molten matter, hardened on its surface by radiation of heat; floats in a a vacuum suspended between the equally powerful, but unequally acting forces of solar repulsion and solar attraction ; is sensitive to the slightest causes tending to disturb its equipoise ; will turn toward the sun without resistance, like any other mass of atoms free to gravitate to an attracting centre, according to its major-axial amount of matter, — now, considering these well-known postulates, none will doubt the sequences, when causes can be shown which will tend to disturb its equipoise.

The positions presented and insisted upon are these: That, when the globe was cool enough to admit the existence of the simpler forms of organized life, it was much larger in all its diameters than it is at this time; that cataclysms which have occurred, and divided time into geological ages, have arisen from sudden disturbances of the earth's equilibrium, its major axis always violently turning toward the sun ; that these events have supervened many times, in consequence of the progressive radiation of heat having contracted the fluid nucleus, making immense voids between the crust and molten mass, into which the crust has suddenly sunk, creating vast basins whereinto the waters have rushed, resulting in new oceans, new plateaus, islands, and mountains, in the disruption of strata, and in a general wreck of Nature, and so disturbing former relations of matter and force as to impart to the globe new equatorial directions ; and that, as a necessary consequence of these events, the shorter axes of the globe — what at present are our poles — are not the result of flattening by rotation, but by a sudden falling-in of surface somewhere, breaking the symmetry of the sphere, and producing an irregular spheroid : from which it follows, as a final corollary, that each great division of time has been ushered in with new equators and new poles, and that the demonstration of Sir Isaac Newton respecting the oblateness of the planet being the result of its rotation is an error, and unworthy of further consideration among geologists.

Physical geography shows the earth to be irregular and unequal in all its latitudes and longitudes. The

7

northern and southern hemispheres are not symmetrical. The antarctic region displays a continent with its subterranean caverns vented by a burning mountain; while the arctic presents a vast depression bounded by abrupt coasts, and containing deep seas which flow between the Atlantic and Pacific. The interior of every continent, and the heart of every sea, smoke with the fire of that everlasting burning, which, ushering in the creation of the planet, proclaims, as eloquently as the voice of God can utter, the processes which have hollowed out its bowels, and fashioned its surface, and fitted it for life and instinct and intelligence. The causes which produce these irregularities and differences lie and act *within* the globe. All know the effects of radiation of heat. It is the same in a planet as in a furnace full of melted rock or metal. Radiation results in crystallization, either regular or amorphous. Heat is convertible into magnetism, and is its equivalent everywhere. If, upon the surface of the globe, we be unconscious of the radiation, from its incandescent nucleus, we are not ignorant of terrestrial magnetism, nor of the necessity of its agency in crystallization. If, then, internal heat assume the form and function of terrestrial magnetism, crystallizing all rocks, or metals and their compounds, pouring its surplus currents into space, and kindling the very fires of the sun itself, we no longer wait for knowledge of the process through which the molten nucleus of the globe shrinks, and by which vast voids insensibly ensue between the nucleus and the crust. Earthquakes which jar the planet from pole to pole, which prostrate cities by instantaneous blows, which raise a thousand

miles of coast a few inches, or a few feet, and let them
down again; volcanoes, the merest bubbles and vents of
bursting worlds, whose eruptions of all sorts are visible
in every hemisphere; terrible detonations, whose bellowings are heard throughout hundreds of leagues of
surface; the simultaneous extinction of Vesuvius with
the submergence of a part of Lisbon; the agitations of
Hecla, and the movements of the waters in the North-
American lakes, on Nov. 1, 1755, — all this class of phenomena indicates and proves the existence of immense
subterranean voids. There is abundant evidence to.
show that they exist in this age. Their capacity we
cannot measure. The present surface of the earth is
comparatively recent. The last great cataclysm is, geo-
logically speaking, not very ancient. Accumulating
evidence compels us to believe, that one of these destructive events has occurred since the human race was
created. The facts I have presented plainly indicate
that another is in the course of preparation. Each
of these vast periodical voids is filled by collapsion of
the surface. The planet must reel under these sudden
and violent transpositions of matter and force, and must
oscillate in its orbit to and fro, before settling into
stable equilibrium, with new equatorial aspects toward
the sun. The major diameter will always preponderate
toward the major diameter of the solar centre. The
minor diameters are never produced, and never have
been produced, by diurnal rotation. The theory on
which this result has been asserted fails when applied
to the form of some other planets. Every theory
which will not cover all well-determined facts is pal-

pably fallacious, and ought to be discarded by exact .
science.

When we examine the general configuration of the
earth's surface, and especially its mountain-systems, it
becomes apparent that their anticlinals can as well result from *depression* of their flanks as from the *upheaval*
of their axes; and, taking this fact in connection with a
constantly condensing nucleus and increasing voids between the crust and nucleus, it is much more reasonable
to suppose the former mechanism to have prevailed
than the latter.

By a more detailed survey of the surface, we observe
minor irregularities, so unequal and unconformable in
various particulars as to reduce the *theory of upheavals* to ABSURDITY; inasmuch as we are compelled to
regard the upheaving force (which is no other than
the molten nucleus) as acting like the irregular discharge of so many batteries of different power beneath
the crust, forcing up one area more than another to
form lesser and greater hills, and, at the same time,
acting as an aggregate, pressing as a unit, beneath entire hemispheres, so as to elevate certain lines many
thousand miles in length, from twenty to fifty thousand
feet high from the bottoms of old and deep oceans.

When this theory is contrasted with that which I am
so bold as to advocate, the same facts will be presented
under new and more natural aspects. Each succeeding
cataclysm, considered as a universal catastrophe, must
leave the globe a wreck, like the ruin of some immense cathedral whose dome and arches have fallen in.
Cornice and frieze, pillar and entablature, broken and

dislocated, lie at all angles of inclination, and in the utmost confusion. So it is with the ancient rocks and more modern strata. Only to this mighty wreck have been added the outgushings of molten matter into fissures, creating dykes, and the unsparing movements of oceans sweeping loose materials and perishing forms of all sorts, from one place to another, partially covering up and disguising the desolation, and softening the general outlines of a new world.

However these remarkable changes may be effected, whether slowly or suddenly, in one way or in another, the equilibrium of the planet must be more or less disturbed, and its tropical relations to the sun must undergo slow or sudden modifications. Gravitation is a constant. Geologists know that the form and features of the planet have undergone frequent changes. If Newton has discovered an eternal and universal law, the splendor of his mathematical authority cannot subvert nor weaken it. The calculus is positive, if it involve no error in quantity nor development. If the spheroidal form of the earth agree approximately with the deductions of mathematicians from mechanical data, *it can only be accidental;* for the rotation of the globe has evidently no more to do with the flattening of its poles than it has in hollowing out an arctic basin, or elevating an antarctic continent. In the days of Newton, nothing was known of geology. Great minds marvelled only at the extent and revelations of the starry heavens. The world was busy in bending all discovery into harmony with the sacred cosmogony. Scientific men have, ever since, been misguided and hampered by sacerdotal ignorance and superstition, and by the arrogance and viciousness of ecclesiastical domination everywhere. It is only of late that geologists have dared to question the real antiquity of the world, or attempted to reduce their facts to science. Even now, men looked upon as great, anointed by kings and senates and universities for the splendor of their abilities, are writing books, distorting facts, and wasting time in endeavoring to reconcile modern science with the sublime fancies of the ancient oriental philosophy. It is impossible to wed the living to the dead, or to create a breathing mastodon from an antediluvian skeleton. When geologists stand wholly aloof from astronomical and theocratical dogmas, apply high physics to their special pursuits, and walk boldy as Nature and Reason lead them, many phenomena will be simply accounted for which have heretofore appeared inexplicable, and beyond the reach of man.

Thus, if we assume that the globe was one hundred or three hundred miles greater in all its diameters when its crust became hard, and was bathed with the earliest seas, and when marine plants and trilobites and mollusca began to appear, the lithological characteristics of the palæozoic ages will be more acceptably deciphered. So, successively, with the carboniferous periods whose vast areas have been folded up and overflowed, and whose fields for reproduction have been so numerous and extensive as to convince us that arctic America, during those remote ages, presented tropical positions to the sun. It will be the same, if we study the coasts of this modern epoch. Their fresh exposures and abrupt aspects, especially in the Pacific Ocean, where I have observed them almost uninterruptedly from 38° north to 14° south latitude, and around Terra-del-Fuego, and along Patagonia, in the Atlantic, indicate sudden violence everywhere. The action of dashing waves and of running water has been over-estimated by geologists. I have observed this widely, long, and carefully. The indications of violent abrasion do not proceed from this cause, but from the breaking-off of continuous areas, which, in vanishing into depths below, leave behind them portions of their wreck in the form of reefs, or precipitous rocks, or sugar-loaf peaks often projecting hundreds of feet above the waves, like the Rock of Gibraltar and Apes' Hill on the opposite side of the strait (the *Columnæ Herculis* of the ancients), which, observed from this point of view, will convince the scientist that they were left standing when the intervening area was submerged, together with the entire bed of the Mediterranean Sea. If the forms, relative positions, and former connections of reefs, rocks, islets, and sugar-loaf peaks in general, be candidly studied by the light of both these theories, — by that of slow upheavals and that of sudden subsidences of surrounding areas, — I have little doubt that the present doctrine of geology will become greatly modified, and that the views I advance will become the leading ones in the science.

When observation is pursued further, and shall embrace certain larger and more isolated points with perpendicular and profound sides, — such as the St. Helenas and Fernando-de-Narhonas in the Atlantic; the St. Lorenzos, Chinchas, and Ballistas, in the Pacific; and the St. Paul's and Amsterdams in the Indian oceans, —

a number of which I have specially noticed or explored, I feel assured that they must be regarded as vestiges of more extensive lands, which, in falling from greater heights, have left these above the water, rather than as primitive masses pushed up from immense depths by such circumscribed and limited displays of force as their form, height, and structure would indicate to have been exerted upon them by present geological theory.

This entire subject, including, as it does, physical geography at large, and geology in its minutest details, requires candid study, and a re-survey from this new and commanding point of view.

At the same time, I wish to state distinctly that I am far from doubting or rejecting the dynamical agency of repulsive forces within the globe, as a means to effect limited changes upon its surface. Trap-dikes, inundations of lava from volcanoes, the upheaval of volcanoes themselves, agitations and fissures of the crust by earthquakes, — all are the results of enormous dynamical power, which acts from the centre to the surface everywhere, alike at the poles and equator, and in every radius of the planet. While this power sustains the roof, or never-ending dome of the planet's crust, it is abortive in all its attempts to rend the globe, and reduce it back to chaotic conditions. Gravitation, the constructive, ruling, and conservative agent, counteracts its destructive tendency by neutralizing its power.

Igneous force may, indeed, be so concentrated, and yet extensive in its action, as to impulsively and transiently uplift a line of coast a few inches or a few feet above its normal level. Molten matter, in circulating

through canals or caverns within the crust (the existence of which I discovered in 1856, on Hawaii, and
described as giving origin to some volcanoes by their
upward bursting from overstrained tension), may, in
consequence of its connection with the central fires,
slowly or suddenly uplift or draw down limited tracts
of surface. I am not unmindful of observations made
upon oscillations of surface, which may be traceable to
dynamical action of this character. But for igneous
force so to act as to upheave that mountain-system of
North and South America, extending from the Icy Cape
to Cape Horn, is a fallacy so monstrous and palpable,
that a mere glance at our physical geography should
lead us to detect and question it. When, however, this
long anticlinal, often indeed extending into double anticlinals, with broad, intervening valleys and basins, is
examined from the point of view upon which this discussion will place the candid physicist, he will discover
it to result from a falling-in of immense lateral exten-
sions, by which the globe has been absolutely reduced
in size. A personal and detailed inspection of the
physical features of the Sierra Nevada, and of the Andes, will finally leave no doubt in his mind of the
correctness of this conclusion.

Further than this, the philosophical geologist will
discover, that the Andes and the Sierra Nevada are only
disfigured remains of vast plateaus, whose present alti-
tudes were, at some antediluvian period, populous with
warm-blooded quadrupeds now extinct; and that all
signs within the range of observation demonstrate the
same physical changes to have transpired there, which

are apparent in lower levels of the globe. There is, therefore, every reason to suppose that the present highest altitude of the Andes was, at one time, the general or lowest level of the surface of the globe, and that the bottom of the Pacific Ocean soared with its lofty ranges of mountains far above them.

If so, then the diameter of the earth, at the poles, must have been very much greater than now, at some more ancient epoch. It must have been more than twenty-seven miles greater to permit such equatorial or tropical exposures to the sun as we know to be necessary for the production of those vegetable forms which abound in the coal measures of arctic latitudes. If it was fifty or a hundred miles greater during any portion of the carboniferous age, it might have been two hundred during the "Taconic" period of the late sagacious and lamented Mr. Emmons, and perhaps three hundred or more, when the life-force began to fashion its primordial and rudimentary organisms upon its waiting surface.

Did the necessities of this discussion demand it, I might illustrate more at length the operation and effects of this mechanism, in producing the multiform aspects of marsh and lowland, plateau and mountain, which constitute our present physical geography, and which have been riven asunder and dislocated in such a manner as to be incompatible with the prevailing philosophy. I will defer, however, elaborate particulars for another occasion; meantime assuming the temerity to predict that *hereafter*, when this branch of inquiry shall have been studied more profoundly, some great geologist will arise, gifted also with the power of numbers, who will measure the earth with rod and plummet, reduce all rocks and strata to their natural order, and reconstruct the globe in all the details of its ancient forms and ages, with their legitimate flora and fauna, and who will bring the system which I have the honor to unfold on this occasion, into perfect harmony with all other departments of natural science.

As a science, this branch of physics is, in itself, purely cosmographical, connecting geology on one hand, with astronomy on the other; and, in its developments, will tend to point out errors, and direct to new paths of discovery in both. It conflicts with truth nowhere. It throws light on problematical questions of much importance. It permits a rational solution of the theories of some illustrious men, which, without its acceptance, can never be harmonized with natural law. It explains the evidences of glacial action anywhere, and in all geological ages, while it rids the theory, as now advocated by its celebrated author, of those incumbrances which must for ever make it untenable. It dispels the darkness which hangs over the speculations of those *savans* who invest the earlier ages with universal and perennial heat, thereby setting at nought those meteorological conditions necessary to tropical fertility. It explains how the antediluvian pachyderms have been violently transported from life and pasture in temperate and tropical regions, to sudden death and immediate congelation by polar cold. It answers all the questions proposed in this thesis, and a thousand more. But, above all, it makes the ways of God upon the earth more clear to man. The veil

that covers the origin and diversity of species is with-
drawn, and the path pointed out, which may yet lead to
the discovery of the transformation of physical into vital
force. Sudden transpositions of matter within and upon
the surface of the globe must necessarily disturb its
equilibrium. The consequent movements of its waters,
and changes of temperature, must equally involve the
destruction of many and the modification of all remain-
ing races. The plan of creation, although mysterious, is,
as we know from its history, progressive. Life is onward
and upward in its instincts and developments. Once
linked to matter, it becomes as unquenchable as the
stars. Every successive condensation, and change of
equipoise, in the globe, have been attended with such
fresh displays of morphological force, that there is now
scarcely a form existing that was originally created upon
its surface. Increasing condensation implies increasing
repulsion. Physical aggregates are composed of molec-
ulars. From the alternate and unceasing play of these
fundamental forces, spring those *secondary* powers of
magnetism, heat, light and electricity, which endow
insensible atoms with motion and sensibility. Con-
sciousness and intelligence, — immortal offshoots of the
Godhead, — the finite within the infinite, cannot exist
independent of these. Thus we catch a glimpse of the
Eternal, rising from matter to force, from force to motion,
from motion to life, from life to intelligence, and from
intelligence to the Creator of all.

In the old geological theory of the world, the zealous
student after truth and progress must plod for ever in an
endless maze. In this which I have attempted to unfold,

the twilight appears to me to be dawning, which shall
lead future generations of scholars to an accurate understanding of nature, and reduce all research after the
origin and phenomena of matter, force, life, and intelligence, into such logical channels, that human reason shall
at last be crowned with a knowledge of the secret spring
of being.

Thus the speculative idea of Leibnitz, announced in
1683, lost for a century, revived by Deluc, and buried
again for several generations, assumes, under our discussions, definite scientific life and force, and crystallizes
around it the accumulating facts and developments of a
new philosophy. Ignorant of the extent of Leibnitz's
conjectures during my own practical explorations toward
the same point, I am only the more confirmed in the
verity and stability of my conclusions as operative and
positive realities, since they are strengthened by the
prophetic inferences of that wonderful mind. Were he
now living, illuminated by the blazing light of modern
discovery, no geologist would be able to resist the
potency of his demonstrations or authority. When first
advanced, the correctness of the opinion could not be
proven; for geology was yet unborn. In the mean time,
human thought has been struggling up among dogmas
and systems, and has lost its way amid the endless confusion of fact and fancy. Even in these ages of increasing light, another century may yet elapse before science
and philosophy will accept, in its fulness, the idea of
Leibnitz, and the unfaltering extension of it presented
in this imperfect disquisition. But, however slow heretofore the progress of knowledge, whatever may be

hereafter the decision of physicists and thinkers, there is one formidable class of public teachers, who, for ever jealous of free thought and inquiry, frowning upon physical discovery, and lagging behind society in all solid acquirements, will not venture to question the possible truth involved in this discussion; for it announces the fulfilment of the alarming judgments which they preach. And well may they, in their turn, truly fear and tremble; for it proclaims, with an eloquence as prophetic as the voice of Peter or Christ, that "the day of the Lord will come as a thief in the night; in the which the heavens shall pass away with a great noise, and the elements shall melt with a fervent heat, and the earth, and the works that are therein, shall be swallowed up."—"But of that day and hour knoweth no man."—"But as the days of Noah were, so shall also the coming of the Son of man be." Nevertheless, the mere *assent* of these opponents of intellectual progress will not prove these predictions to be correct, any more than their tortures, rebukes, and discouragements have proved the discoveries of Galileo or Darwin to be wrong. The final triumph of scientific truth will only put its enemies and cavillers to the deeper shame.

To the really devout student of Nature, who loves and cultivates science because it leads to the surest knowledge of that Invisible Being whose marvellous counsels are undivided, it will be pleasant to perceive that a remarkable correspondence exists between the physical changes herein delineated and the events and mechanism as succinctly described in the sacred cosmogony. In no other manner could those events as there recorded be fulfilled : " Let the waters under the heaven be gathered together unto one place, and let the dry land appear ; and IT WAS so." How could it have been " so," unless the arch of the globe collapsed, making a great continuous basin, wherein the waters were gathered together, thus causing continents to appear? The idea of Moses is precise, and it has no relation nor pointing to theories of upheaval. It does not mean, Let the dry land be upheaved, and all the waters under the heavens be displaced, and old lands be overwhelmed; but " Let the waters be gathered together" into a deep place, so that new lands may appear above them. Nothing expressed in language can be more definite.

Thus we see, that by a fearless pursuit of natural truth, discarding all reverence for the worthless systems of sophists, ecclesiastics, and theorists, we shall meet with the greater satisfaction in the end. The end, indeed, we may be slow to reach ; but patient labor, self-sacrificing travel, extensive observation illuminated by the largest culture, candid comparisons of facts from every theoretical point of view, broad generalizations based upon rationalism rather than upon mathematics, — in a word, fearless, able, and honest inquiry, following the lead of Nature with the simplicity of a child in the hand of his father, will, at last, raise Geology to the high rank it ought to occupy, and establish it upon solid foundations as an EXACT SCIENCE.